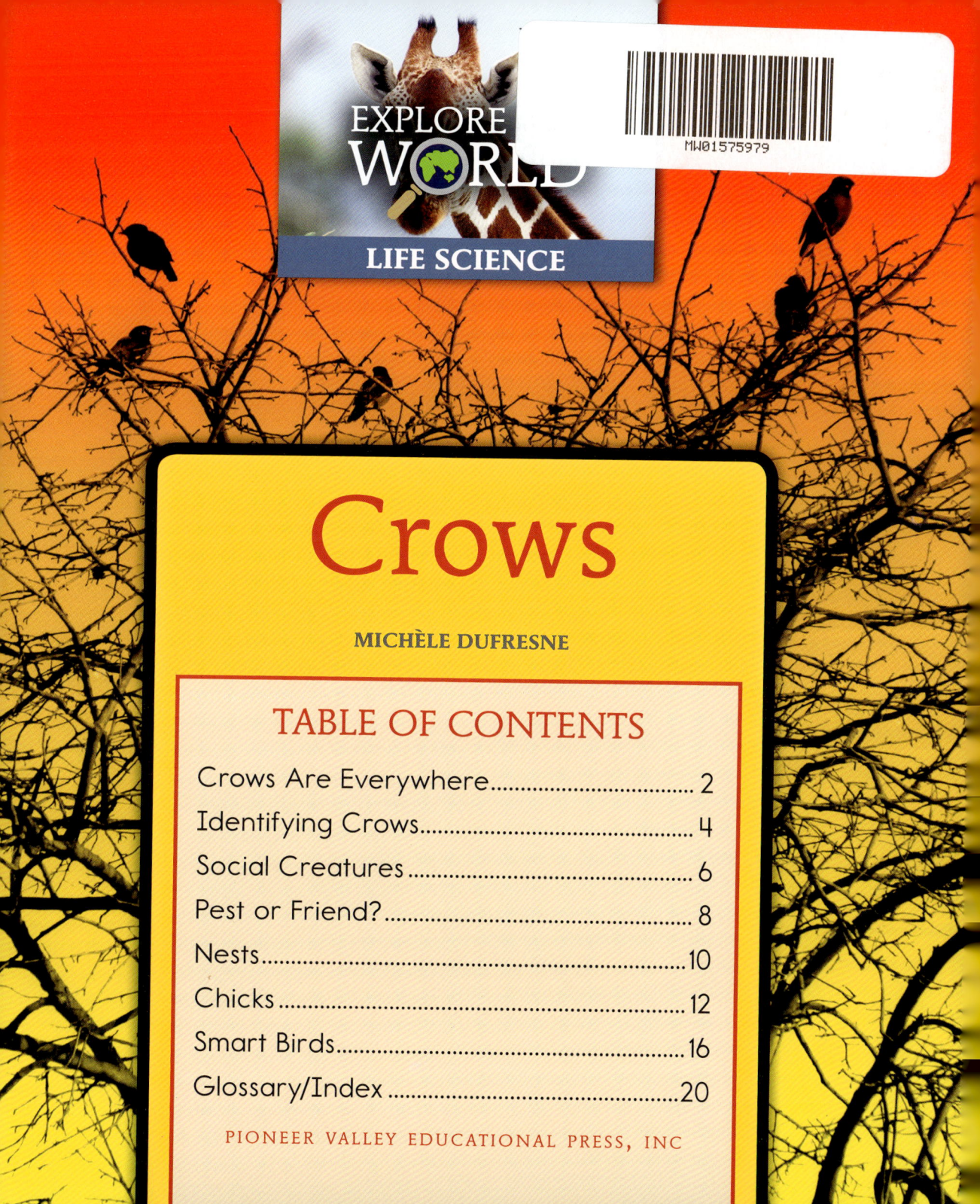

Explore the WORLD
LIFE SCIENCE

Crows

MICHÈLE DUFRESNE

TABLE OF CONTENTS

Crows Are Everywhere	2
Identifying Crows	4
Social Creatures	6
Pest or Friend?	8
Nests	10
Chicks	12
Smart Birds	16
Glossary/Index	20

PIONEER VALLEY EDUCATIONAL PRESS, INC

CROWS ARE EVERYWHERE

Crows are large black birds.
You often see crows in fields,
at the tops of trees, and in parks.
If you listen, you may hear
their loud cawing voices
calling to each other.
Crows are everywhere!

INDENTIFYING CROWS

Crows have **glossy** black **plumage** that can sometimes look purple in the sunlight.

Their beak is also black and has a slight hook at the end.

Male and female crows look alike.

square-cut tail feathers

SOCIAL CREATURES

Crows are very **social**. They like to be around other crows. Sometimes millions of crows will flock together.

Crows are also **aggressive**. They will work together to harass or drive off a **predator**. They can chase away larger birds, including hawks and owls.

>> Crows usually post guards who alert other crows of danger.

PEST OR FRIEND?

Crows live near fields, woodlands, and forests.
They also like to live near people,
where they can find food.
You will find crows in farm fields,
parking lots, backyards,
along the side of the road,
and in garbage dumps.

People who plant gardens sometimes put a scarecrow in it to frighten crows away.

Crows will eat almost anything including insects, earthworms, fish, snakes, eggs, mice, fruits, and vegetables. They also eat dead animals that have been run over by a car. Crows will even look for food in garbage cans!

Crows can damage crops by eating the plant before it is ready to be picked. Crows can also help farmers by eating insects that can attack their crops.

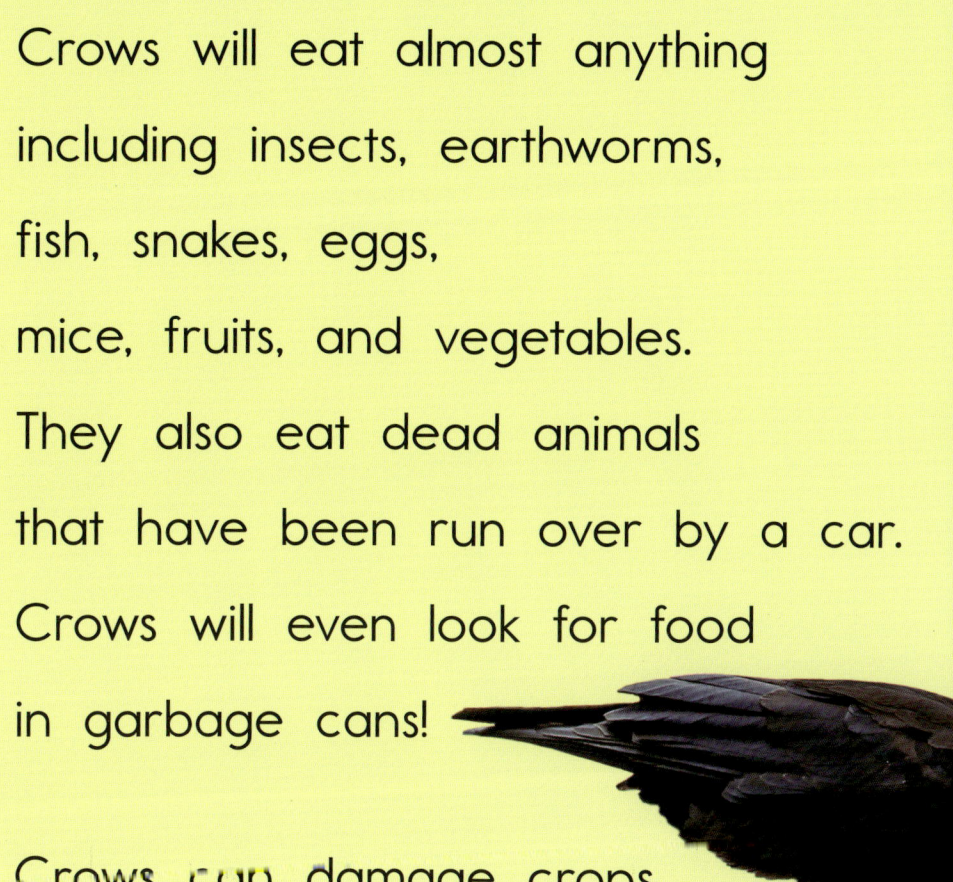

NESTS

Crows like to build their nests in **lofty** places.

Their nests are often found near the tops of tall trees.

In places where there are no tall trees, crows may build their nests in bushes.

In cities, crows may even build their nests on window **ledges**.

MORE TO EXPLORE

Crows build their nests out of branches and twigs that they line with bark, plants, hair, twine, cloth, and other soft material.

MORE TO EXPLORE

Other birds and animals such as raccoons and squirrels sometimes use a crow's old nest as a summer napping place.

CHICKS

Male and female crows
build the nest together.
When the nest is completed,
the mother crow lays
four to five eggs.
She sits on her eggs for 18 days.
Her mate will bring her food to eat
so she can stay at the nest
to keep their eggs warm.

After the eggs have hatched,
the parents will make a lot of noise
if any bird, animal, or person
comes near the nest.
The young chicks also make
a lot of noise and can be heard
cawing at their parents to give them food.

The chicks grow quickly. They leave the nest about four weeks after hatching, but their parents will still feed them for another 30 days.

MORE TO EXPLORE

During late summer, fall, and winter, crows gather from many miles around to form large groups that **ROOST** together.

15

SMART BIRDS

For a long time, people thought that only humans made and used tools. Now we know that some crows make and use tools, too!

Crows will use a stick to dig **grubs** out of logs and branches.

Crows have been seen dropping nuts onto a street, and then waiting for passing cars to crack them open. Then they swoop down to eat the nut that is inside the shell.

Crows are so smart they can "talk" to each other. They change their "caw" sound to warn each other about danger. Other animals that hear the warning understand that a dangerous predator is nearby.

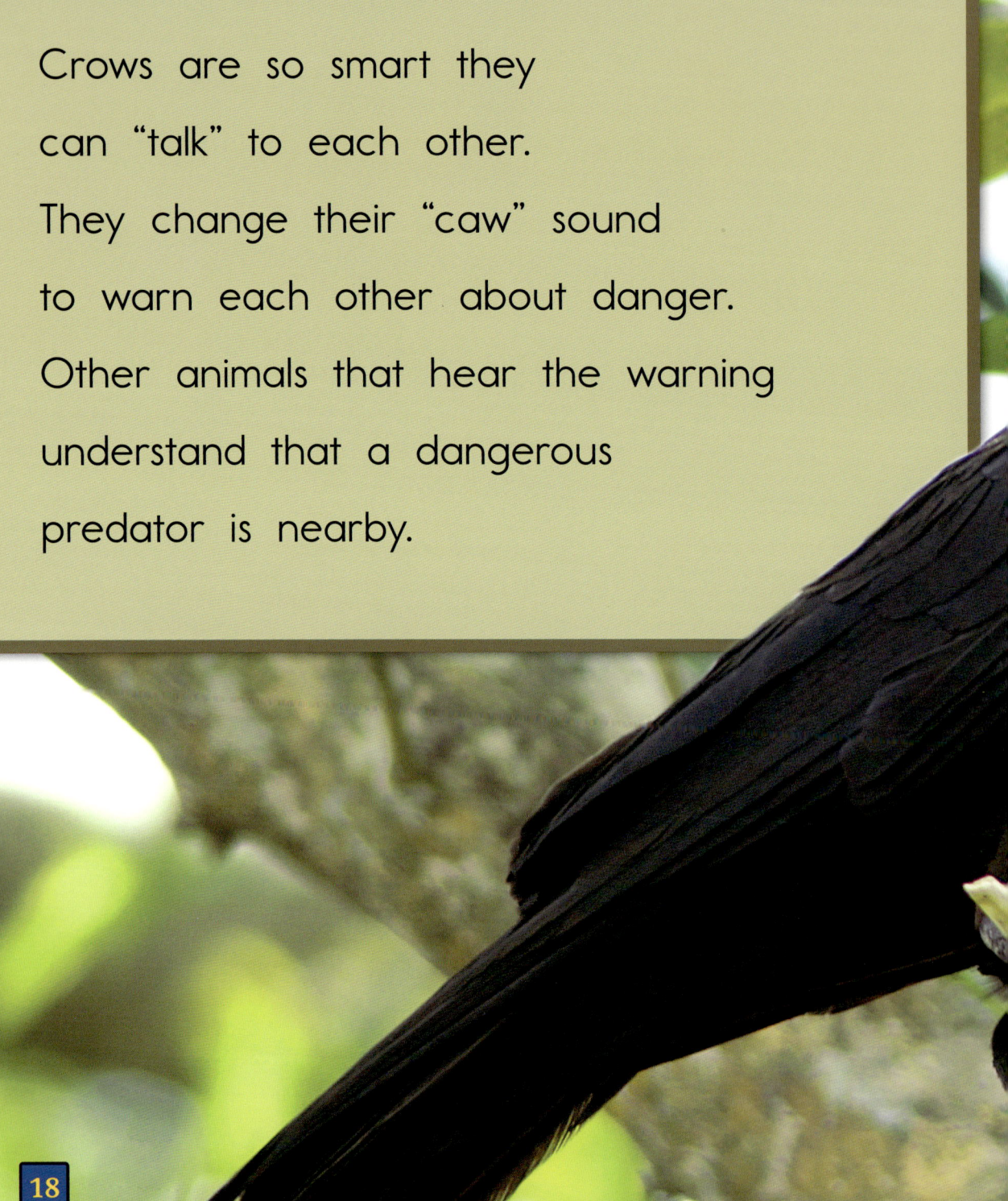

Crows
Myths and Literature

Crows are found in many stories around the world.
Read this familiar story about a crow.

Once upon a time there was a crow who was very thirsty. He saw a pitcher with only a little bit of water in it. He tried to drink the water but his beak was not long enough to reach it.

INDEX

aggressive 6
cawing 2, 13, 18
chicks 12–13, 14
crops 9
eggs 9, 12–13
feathers 4–5
female 4, 12
flock 6
food 8–9, 12–13
garbage 8
grubs 16
guards 7
hawks 6
nests 10–11, 12–13, 14
owls 6
predator 6, 18
roost 15
scarecrow 8
social 6
tools 16–17
warning 18

20

GLOSSARY

aggressive
ready to fight

glossy
having a shiny, smooth surface

grubs
young forms of insects that look like worms

ledges
narrow flat surfaces that stick out from a wall

lofty
very tall

plumage
the feathers that cover a bird's body

predator
an animal that kills and eats another animal for food

roost
a place where birds rest at night

social
likes to be with others

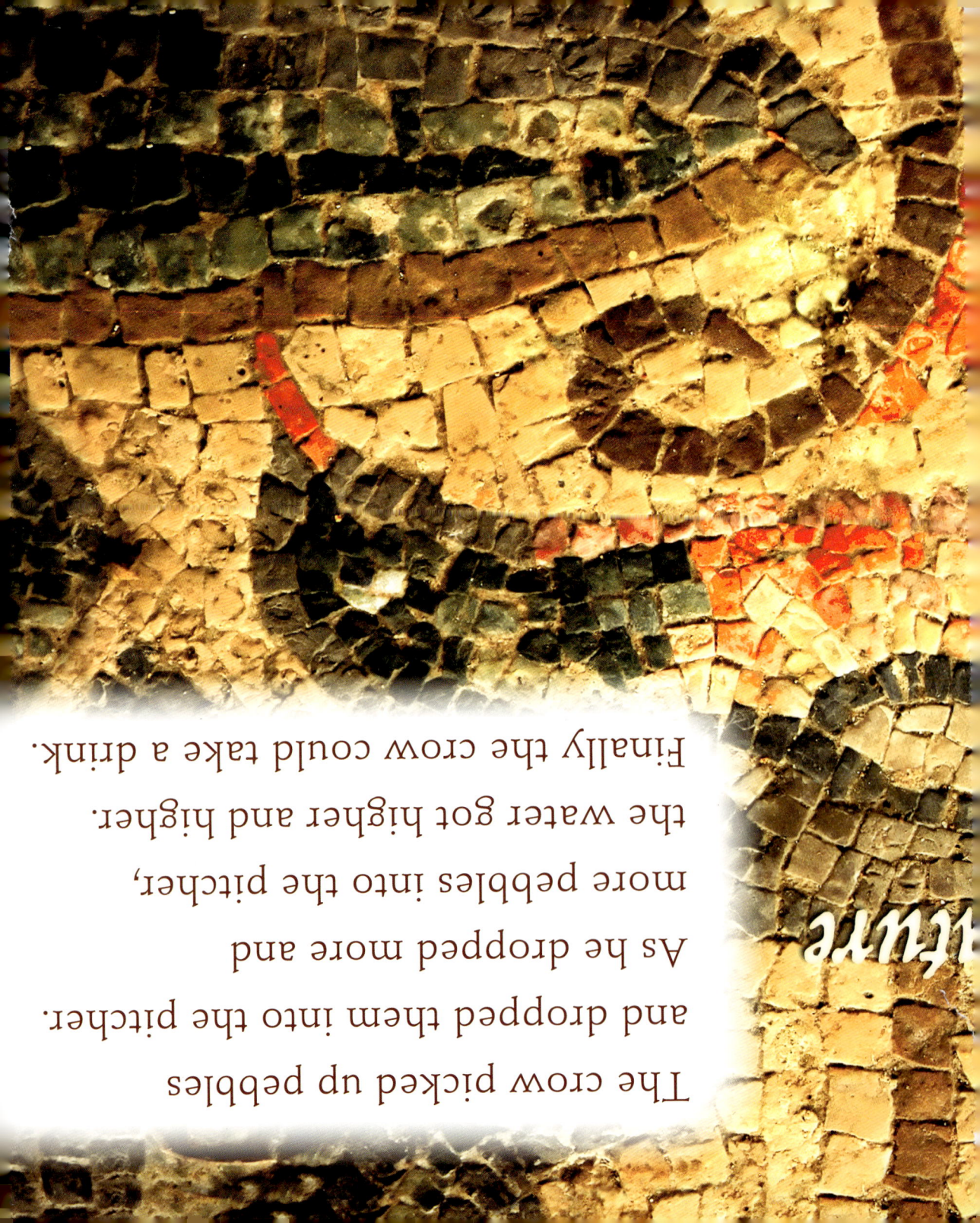

The crow picked up pebbles and dropped them into the pitcher. As he dropped more and more pebbles into the pitcher, the water got higher and higher. Finally the crow could take a drink.